It's a Numbers Game

By Jane Kelley

Contents

Introduction

Sports competitions are a big part of our lives. Some people enjoy playing sports such as soccer. Others like attending sports events, and some enjoy reading about sports or watching them on TV. No matter how people choose to participate, when major competitions such as a World Cup or the Olympic Games happen, almost everybody discusses the teams and **athletes** who are in the spotlight. Who won? Who lost? Who improved? Who collapsed right before the finish?

Statistics are a great way to make sense of sports. Yet, what exactly are they? In simple language, statistics are collections of numbers that help people understand something.

Who's Really Number One?

Many people believe that David Beckham is the greatest soccer player ever. Others believe that Pelé is. What will the numbers say? It's true that Beckham is one of the greatest players to ever play soccer. He has played in 644 games and scored 119 goals. However, Lionel Messi of Argentina has played only 233 games, yet he's scored 150 goals. Pelé has played in 1,360 games and has scored 1,280 goals.

◀ David Beckham

Lionel Messi ▶

Is there a way to compare the number of goals these players have scored even though they haven't played in the same number of games? Yes, there is. We can fairly compare the three athletes' accomplishments with the help of statistics.

Pelé ▶

Let's look at those numbers again, but this time in a table:

Player	Goals	Games
Pelé	1,280	1,360
Beckham	119	644
Messi	150	233

We can use these numbers to work out each player's scoring average per game. In sport, the average is figured out by dividing the player's scores by their total chances to play. By working out the scoring average for each soccer player, we can compare their performance and see who might be the best.

If we want to find Pelé's scoring average, we find out the total number of goals he scored compared with how many chances he had to score goals. In soccer, this means how many games he played. We divide 1,280 by 1,360. That equals 0.94. In other words, on average, Pelé scored 0.94 goals in each game he played. On average, he scored almost a goal every game!

We can find the scoring averages of Beckham and Messi in the same way.

Player	Goals	Games	Scoring average
Pelé	1,280	1,360	.94
Beckham	119	644	.18
Messi	150	233	.64

When you compare the scoring averages, there are big differences. But goal scoring is just one part of soccer. Beckham could still be your favorite, because there are other qualities that can make someone a great soccer player.

When two teams play each other, it's pretty easy to see which team is better. The one that scores the most points wins. Still, what if you want to find out which team is the best team *ever*?

We can use **percentages** to help us compare performances of teams. Percentages can tell us how often something happens (such as how often a team wins). Percentages are always shown as a **fraction** out of 100. For example, 10 percent is another way of saying $^{10}/_{100}$ or 0.1. Percent can also be shown using %, so 10 percent can be written as 10%.

A score of 100% means that the "something" (such as the team winning) *always* happens. A score of 0% means that the "something" *never* happens.

This chart compares five past NBA basketball teams with the most wins in a season.

Team	Wins	Losses	Winning percentage
Chicago Bulls 1995–1996	72	10	87.8%
Los Angeles Lakers 1971–1972	69	13	84.1%
Chicago Bulls 1996–1997	69	13	84.1%
Philadelphia 76ers 1966–1967	68	13	83.9%
Boston Celtics 1972–1973	68	14	82.9%

How do we find the winning percentage for each of these teams?

First, find the total number of games that each team played For the Chicago Bulls' 1995–1996 season, we add 72 wins and 10 losses to get 82 games. What part or percent of 82 is 72? Just like before, $^{72}/_{82}$ is the same as saying 72 divided by 82, which is .878. Because we want the percentage, multiply the result by 100 to get 87.8 percent. From 1995–1996, the Chicago Bulls won 87.8 percent of their games, which is a pretty impressive statistic!

If you keep track of a team's wins and losses, you can see how a team is measuring up when they have only played part of their season. You can also compare teams from different years. Using numbers, you can compare athletes' times or distances even if they never competed directly against each other.

It's not just on the field, the court, or the track that statistics are important. The water is another place you'll find numbers – and lots of them!

Ian Thorpe was born in Australia in 1982. He started swimming when he was young. He was allergic to the **chlorine**, so he swam his first race with his head out of the water the whole time. He looked kind of odd, but he won. Pretty soon he outgrew those allergies and kept on winning. In 2001, he became the first Australian swimmer since 1959 to win medals in *every* race length from 100 meters to 800 meters.

Ian Thorpe wins races because he works hard, and he has a powerful kick that helps him speed up while he swims. Many people believe that his size-17 feet act as flippers to help him slice through the water.

◄ Ian Thorpe is known as "The Thorpedo" because he destroys records in the pool.

FASTEST PARALYMPIAN

Oscar Pistorius had both of his legs **amputated** as a child. With the help of his Flex-Foot Cheetah®, though, he can run 100 meters in 10.9 seconds – that's just 1.32 seconds slower than Usain Bolt's world-record time.

Competitive swimming has been around for a long time. Records are always getting broken. In 1908, Henry Taylor won a 400-meter freestyle race, and broke the world record. His time was 5 minutes, 36.8 seconds. By using a graph, we can compare the speeds of other world-record holders since then.

◄ Henry Taylor (left)

In the 1920s, Johnny Weissmuller won ► five Olympic gold medals in swimming.

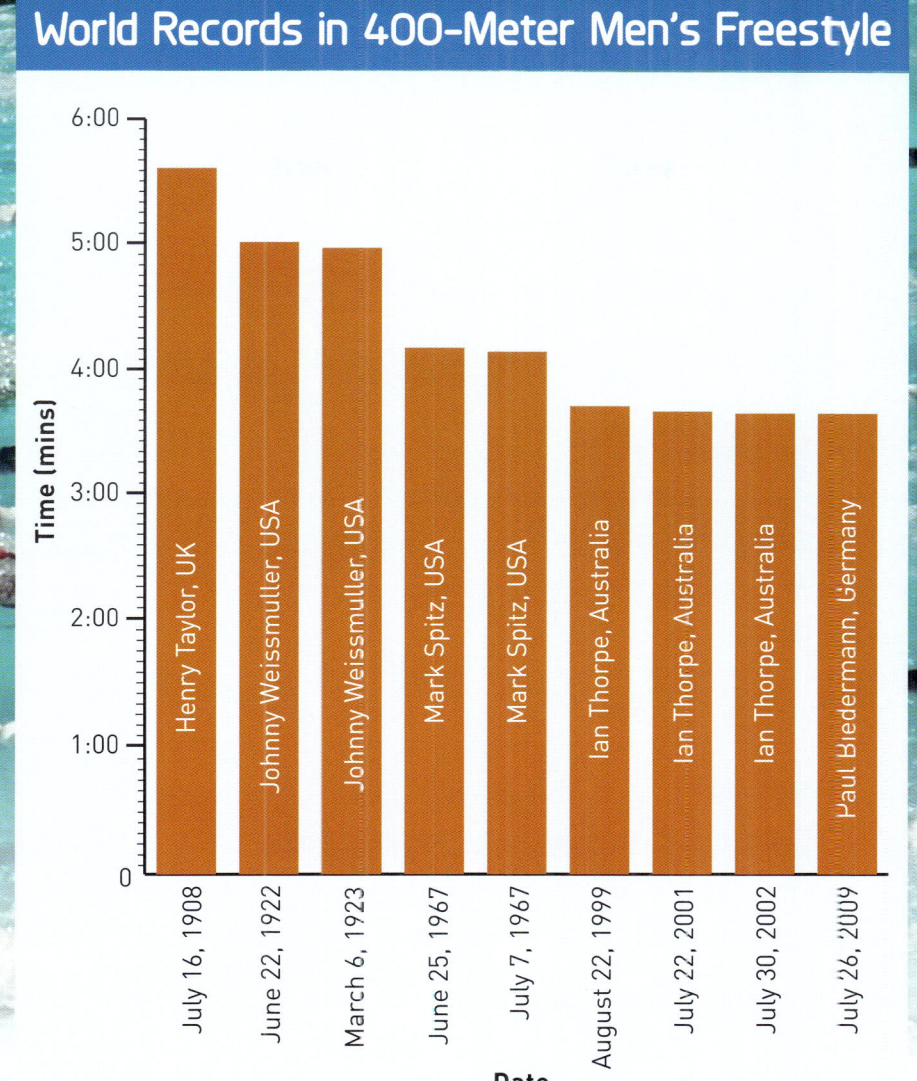

World Records in 400-Meter Men's Freestyle

Time (mins)

- Henry Taylor, UK — July 16, 1908
- Johnny Weissmuller, USA — June 22, 1922
- Johnny Weissmuller, USA — March 6, 1923
- Mark Spitz, USA — June 25, 1967
- Mark Spitz, USA — July 7, 1967
- Ian Thorpe, Australia — August 22, 1999
- Ian Thorpe, Australia — July 22, 2001
- Ian Thorpe, Australia — July 30, 2002
- Paul Biedermann, Germany — July 26, 2009

Date

As you can see, swimming records are often getting broken. Can you guess how fast swimmers in the future will be able to swim the 400-meter freestyle?

Way above Average

Sports where athletes don't score goals or compete against the clock also use statistics. These sports almost always use numbers to determine the winner.

In competitions for sports like gymnastics, judges use numbers to score each participant. Rules are in place to make the scoring as fair as possible. Each judge watches the event and decides how many points to give each competitor for each routine they do. Yet, since each judge's view is different, is there a fair way to decide how many points to give out?

During the 1976 Olympics in Montreal, Canada, Japanese gymnast Shun Fujimoto was performing his floor exercise routine. Near the end of a tumbling run, he somersaulted and twisted his knee the wrong way when he landed. He felt a jolt of pain, but didn't tell anyone that he was injured. He went on to compete in his next event – the **pommel horse**.

Even though he was suffering, he kept his concentration and did very well. The judges gave him a score of 9.5 points out of a possible 10. Just where exactly did that 9.5 come from?

SHUN FUJIMOTO BROKE HIS KNEE DURING HIS TUMBLING ROUTINE AT THE 1976 OLYMPICS.

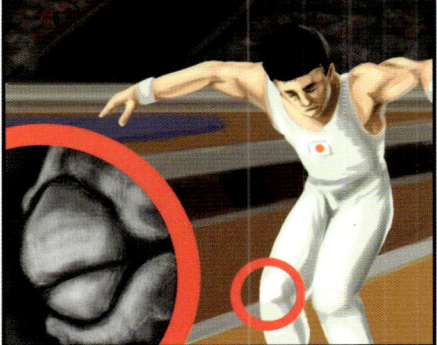

HE DIDN'T TELL HIS TEAMMATES, HOWEVER, WHO WERE FOCUSED ON BEATING THE RUSSIAN TEAM – FOR A GOLD MEDAL.

SHUN DIDN'T WANT TO LET HIS TEAM DOWN, SO HE WENT ON TO COMPETE IN THE POMMEL HORSE ...

AND THE RINGS, WITH A BROKEN KNEE. HE KNEW THE FINAL DISMOUNT FROM THE RINGS WOULD BE PAINFUL ...

AND IT WAS ... WHEN HE LANDED, HE **DISLOCATED** HIS ALREADY BROKEN KNEE.

BUT HE DIDN'T SHOW HIS PAIN. HE EVEN SMILED AT THE JUDGES.

SOON AFTER, THOUGH, HE COLLAPSED IN AGONY. YET WITH HIS SCORE ADDED TO THE TEAM SCORE, HE HELPED HIS TEAM WIN THE GOLD MEDAL, BY A MERE 0.4 OF A POINT.

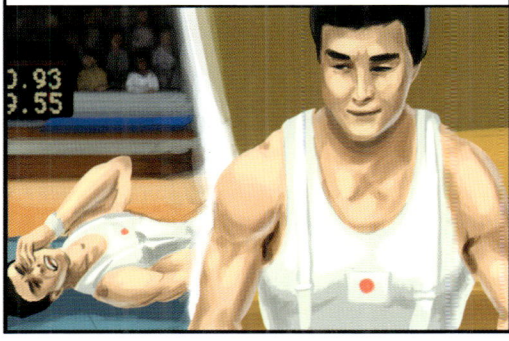

Six judges watched Fujimoto swing his legs over and around the pommel horse. They kept track of small errors in the performance and made **deductions**. Since judging is **subjective**, every judge is different. They all try their best to be fair, but some see what others don't. Since not every judge catches the same mistakes, each gymnastic **apparatus** event has a panel of six judges.

So how do the athletes end up with only one score?

The judges' scores are **averaged**. In many competitions, the highest score and the lowest score are ignored. The remaining four scores are added together and the total is divided by 4.

What would happen if the highest score and the lowest score were kept?

Let's say that a gymnast received scores of 9.0, 6.2, 8.6, 8.5, 7.9, and 8.8. This is really unlikely in gymnastics, but it will help you see how the scores work.

Add those scores together and you get a total of 49.0. Divide that by 6 (because we're using 6 scores). The answer, which is the average, is 8.16.

What's the average if we take out the highest score of 9.0 and the lowest score of 6.2? Add 8.6, 8.5, 7.9, and 8.8 to get 33.8. Divide that by 4 (because we're using 4 numbers this time). Now the gymnast's average score will be 8.45. The difference of 0.29 (or 0.3) doesn't seem very big, but in competitions, 0.3 often means a lot.

The pommel horse is a men-only event. Points are deducted if the gymnast brushes or hits the apparatus with his legs or body.

In the 1976 Olympics, the Japanese gymnastics team was hoping to beat the Russian gymnastics team for the first time ever. That was why Fujimoto kept competing even *after* his injury. His score of 9.5 on the pommel horse and 9.7 on the rings were added to his teammates' scores. The Japanese gymnasts got a total score of 576.85. The Russians got 576.45. The Japanese had beaten the Russians by only 0.4 of a point. After the competition, doctors told Fujimoto that his kneecap was both broken and dislocated.

Fujimoto received a gold medal for the team event. He also earned the gratitude of his country and the respect of many because he conquered his pain to help his team finally win Olympic gold.

FIRST PERFECT 10

Romanian gymnast Nadia Comaneci was only 14 years old when she received a perfect 10 score from the judges during the 1976 Olympic Games. It was the first perfect score ever achieved in gymnastics at the Games. The score showed up as 1.00 on the scoreboard, since there was only space for three digits. Nadia went on to achieve seven more perfect scores.

Turning Heads

Olympic gymnasts aren't the only athletes who perform amazing displays of athletic abilities. Many who watched the 2010 Winter Olympics were blown away by the snowboarders. Judges scored these competitors, too, by keeping track of how they did on their 360, 270, and 540 jumps and twists.

MOST IDITAROD WINS

Susan Butcher won the 1,150-mile (1,851-kilometer) **Iditarod** dogsled race four times: in 1986, 1987, 1988, and 1990. Since the race started in 1973, fewer than 10 percent of the winners have been women. (Of course, in reality, dogs have won every single time!)

Shaun White was one of the favorite snowboarders in the 2010 Winter Olympics. When he was born in 1986, it might have been difficult to imagine him becoming an Olympic champion. Because of a **heart defect**, he had to have two open-heart surgeries before the age of one. However, his illness didn't slow him down for long. By the time he was nine, he was getting a lot of attention for his advanced skateboarding skills. So when he tried snowboarding, no one was surprised that he had huge success. In the 2006 Winter Olympics, he won a gold medal for the half-pipe in snowboarding. Four years later, during the 2010 Winter Olympics half-pipe competition, Shaun had received the highest score before his turn in the finals. He didn't even need to ride again to get the gold, but he competed in the finals anyway. It was a good thing he did, because he landed a very difficult Double McTwist 1260.

◀ Shaun White

To most people watching, Shaun probably seemed like a blur as he spun up over the half-pipe course. However, the judges watched carefully to see if he had done the trick. The number "1260" is very important. What does it mean?

Every circle can be divided into small segments called degrees, and each circle has 360 degrees.

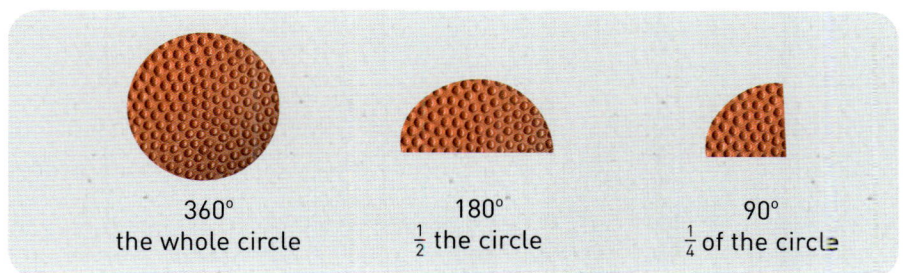

| 360° | 180° | 90° |
| the whole circle | $\frac{1}{2}$ the circle | $\frac{1}{4}$ of the circle |

When anything **pivots** around a whole circle, it has turned 360 degrees. If it only turns half of the circle, then it has turned half of 360, or 180 degrees. If it turns one-quarter of the circle, then it has turned 90 degrees.

The usual McTwist is 540 degrees. How many times would someone have to spin, or rotate, to reach that number? Adding 360 and 180 equals 540. To do a McTwist, the snowboarder would have to go one full circle (360) plus one half circle (180), or rotate $1\frac{1}{2}$ times.

Shaun did much more than that. He did a Double McTwist 1260. Can you figure out how many rotations that is? Adding 360 and 360 equals 720, or twice around, and 1260 is more than that: 360 plus 360 plus 360 plus 180 equals 1260. Shaun rotated $3\frac{1}{2}$ times around!

Snowboarding isn't the only board sport that scores competitors by spins. Skateboarders and surfers also do tricks that involve twisting and spinning in the air.

Scores or Saves?

Statistics don't just tell us who is the best. These numbers can also provide valuable information to help players improve their games. No team can win just by scoring points. Teams also have to have a defense to keep their opponents from scoring.

In sports such as ice hockey or lacrosse, one player guards the goal and blocks the puck or the ball so it doesn't go into the net. Another very important sports statistic involves knowing exactly how many times a goalie has let the other team score. This statistic is called the goals-against average.

The goals against the goalie, or the number of goals made by opponents, are added up. Then, this total is divided by the total number of games played.

Another defense statistic that is used in games such as soccer, ice hockey, and lacrosse is the save percentage. What's the difference between the goals-against average and the save percentage?

The goals-against average shows the average number of times the other team scores. The save percentage shows the percentage of times the other team *tries* to score but doesn't.

To find the save percentage, divide the number of shots that the goalie saved by the number of shots that the other team made at the goal. This will give you a fraction that you can convert into a percentage.

Tim Howard was born in
New Jersey in 1979. When
Tim was about ten years
old, he learned he had
Tourette syndrome. This is a
neurological condition that
causes people to twitch or
jerk or sometimes shout out
strange words. Tim refused to take
medication for his Tourette syndrome
because he thought it would slow him down. Instead, he
worked very hard to control the condition himself.

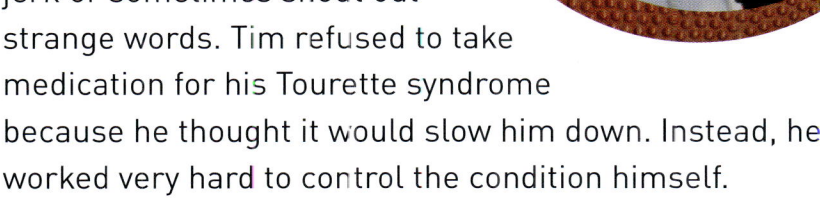

Today, Tim is one of the best goalkeepers in the world. Here
is a chart of his 2007–2010 statistics while he played for
Everton, which is one of the top soccer teams in England.

Year	Games played	Goals allowed	Goals-against average	Shots on goal	Save percentage
2007	36	30	**.83**	169	**82.2%**
2008	38	37	**.97**	195	**81.0%**
2009	38	49	**1.29**	169	**71.0%**
2010	15	19	**1.27**	55	**65.5%**

As you see, in 2007 Tim allowed only 30 goals in 36 games.
To find the average number of goals he allowed in each
game, divide the total number of goals by the number of
games played.

To get the save percentage you start with the shots on goal, or the shots that the opposing team made on target at the goal. In the 2007 season it was 169. Subtract the number of goals: 169 minus 30 equals 139. Divide this number by the number of shots on goal: 139 divided by 169 equals about 82 percent. That's the percentage of how many of the shots on goal Tim was able to keep from scoring.

If you look at the chart, you notice that there were only records available for part of 2010. Yet, even though Tim had played fewer games and faced fewer shots, the statistics can be compared to previous years.

You will see that his goals-against average increased, while his save percentage decreased. Does this make sense? Yes. If opponents score more, then he saves fewer. Now if only the numbers could tell Tim how to keep his opponents from scoring!

Statistics in Action

With numbers, we can always find the winner, compare athletes, and learn what we need to do in order to improve our athletic performance.

Of course, just because it is possible to find a number, that doesn't mean that number is always useful. Maybe the coach of your team has kept track of when you win on odd-numbered days and even-numbered days. Those numbers would be called random, since they happen by chance or without reason. However, if your team wins twice as many games away as they do at home, the coach can use that information to help your team do better on their own **turf**.

With statistics, the speed of Ian Thorpe, the spins of Shaun White, and the courage of Shun Fujimoto can all be turned into numbers. Those numbers can both inspire and teach us. Of course, sometimes they are just fun to know!

Glossary

amputated – cut off, usually by surgery

apparatus[11] – piece of equipment

athletes[6] – people who are good at, and practice, exercise or sports

averaged – added together and divided by the total number of scores

chlorine – a chemical used to purify water in swimming pools so people don't get sick when swimming

deductions – things that are taken away or subtracted

dislocated – put out of place; moved from its normal place

fraction[2] – a quantity less than a whole, or less than one

heart defect – a problem in the heart, usually from when a person is born

Iditarod – a dogsled race that is 1,150 miles (1,851 kilometers) across Alaska – from Anchorage to Nome – and that takes between ten and seventeen days

Māori – the native people of New Zealand

neurological – related to the nervous system

percentages[2] – parts of the whole; for example, ¾ equals 75 percent (%) of 1

pivots – turns around a point

pommel horse – a padded "block" on legs with handles on the top, used for men's gymnastics

subjective – seen differently by every person

turf – ground; "own turf" means an area that you feel is your own, where you are comfortable; in sports, it means playing in the same place or same city as you usually practice in

Academic Vocabulary Key	4	Economics	8	US History	12	Technology
1 English Language Arts	5	Civics	9	World History	13	General Arts
2 Mathematics	6	Geography	10	Health	14	Dance/Music
3 Science	7	General History	11	Physical Education	15	Theater/Visual Arts